Peru Public Library
Peru, Indiana 46970

PERU PUBLIC LIBRARY
3 3069 00037154 8
Rabinowich, Ellen/Kangaroos, koalas, and
590 RABINOWICH C.1 ADULT

599

238123

Rabinowich, Ellen
    Kangaroos,
Koalas and other marsupials

Peru Public Library
Peru, Indiana 46970

Peru Public Library
Peru, Indiana 46970

Kangaroos, Koalas, and Other Marsupials

KANGAROOS, KOALAS, AND OTHER MARSUPIALS

by Ellen Rabinowich

FRANKLIN WATTS
NEW YORK | LONDON | 1978
A FIRST BOOK

FOR MY PARENTS

Cover photographs courtesy of The Australian Tourist Commission

Photographs courtesy of: Australian Tourist Commission: pp. 1, 2, 15, 20, 30, 36, 58, 60, 62, 68, 72; Australian Information Service: pp. 5, 6, 19, 23, 24, 29, 34, 49, 50, 53, 55, 57, 61, 65, 67, 71, 73, 75; Zoological Society of London: pp. 8, 16, 44, 45, 46; San Diego Zoo/Garrison: pp. 12, 17, 38, 39, 42, 66, 76, 79; Annan Photo Features/Thomas: p. 22; Los Angeles Zoo/Sy Oskeroff: p. 27; Photo Trends/John Drysdale, Camera Press, London: p. 32.

Library of Congress Cataloging in Publication Data
Rabinowich, Ellen.
    Kangaroos, koalas, and other marsupials.
    (A First book)
    Bibliography: p.
    Includes index.
    SUMMARY: An introduction to the characteristics, habits, and habitats of various marsupials, the animals with pouches for their young.
    1. Marsupialia—Juvenile literature. [1. Marsupials] I. Title.
QL737.M3R3   599'.2   78-5805
ISBN 0-531-01489-4

Copyright © 1978 by Ellen Rabinowich
All rights reserved
Printed in the United States of America
5 4 3 2

Contents

WHAT IS A MARSUPIAL?   1

THE AMAZING KANGAROO   6
Birth   10
The Pouch   11
Growing Up   13
The Teeth   18
The Big Feet   21
The Long Tail   23
The Hind Legs   23
Mobs   25
Mob Life-styles   26
Courtship   28
Breeding   28
Wrestling   31
Boxing   31
Big Reds   35
Great Gray   37
Euro   40

Tree Kangaroos	41
Prettyface Wallaby	43
Nail-Tails and Other Wallabies	43
Rat Kangaroos	47
Dingo	48
The Kangaroo Fights Back	48
People	51
Sheep Ranchers	52
Motorists	54
Big Business	54
KOALAS	**57**
Bringing Up Baby	64
Strange Koala Sounds	69
Koala Pets	69
The Koala's Enemies	70
A Second Chance	70
OTHER MARSUPIALS	**73**
Marsupial Moles	74
Opossums	77
FURTHER READING	**81**
INDEX	**83**

Kangaroos, Koalas, and Other Marsupials

Peru Public Library
Peru, Indiana 46970

The cuddly koala is one of Australia's many marsupials.

Have you ever seen a baby kangaroo peeking out from its mother's pouch? Many people believe that only kangaroo mothers carry their babies in this way. This is untrue. Other animals have pouches. They are called *marsupials,* or pouched mammals. Marsupial comes from the Latin word *marsupium,* which means "pouch."

The koala, which looks like a teddy bear, is also a marsupial. So is the sugar glider, a furry creature that loves sweets and sails through the air without wings. There are nearly 175 species, or kinds, of marsupials in Australia, New Guinea, and the islands nearby. Only one species of marsupials lives far from the rest— the opossum. This animal lives in North and South America.

All marsupials are mammals. They are warm blooded, covered with fur, and the young are nourished by milk from the mother's mammary glands. However, marsupials are less advanced than other mammals like horses and apes because marsupials do not have a placenta. The placenta is a structure inside the mother's belly through which the baby obtains food. Babies attached to placentas grow and develop inside the mother's body and are born fully formed. The newborn are much smaller than their parents, but they still have eyes, a nose, and ears. Marsupial newborn look nothing like their parents. After birth they develop and grow in a pouch outside their mother's belly.

Why do all marsupials except the opossum live only in the Australian region? Scientists have to look a long way back to find the answer—between 65–70 million years, in fact. At that time, the earth was very different from what it is today. Instead of separate continents, some lands were joined together. For example, the continents we know as America, Antarctica, and Australia were connected. Different theories have been put forth about how marsupials reached Australia, but today most scientists believe that marsupials originated in North America, went to South America, and crossed through Antarctica to reach Australia. No matter how they got there, one thing is certain. After they arrived, the earth underwent enormous changes. Mountains erupted, glaciers melted, and oceans flowed between lands. One of the results of this was that Australia split or drifted away from Antarctica. With water surrounding them on all sides, the marsupials couldn't go back to America, even if they wanted to. They were cut off from the rest of the world.

This was very fortunate for the marsupials. If they had remained in America, they would have had to compete for food and living space with the more advanced placental mammals. In fact, except for the opossum, the marsupials that remained in America eventually died out. But those that went to the Australian region had nearly the whole space for themselves. Because the terrain was varied, they adapted with various life-styles. Some burrowed into the earth. Others made their homes in trees. And many lived in the open grasslands and rocky hills. The environment suited them so well that after millions of years they have managed to survive in much the same form.

Naturally, over such a long period, some kinds of marsupials died out. But, today, many still exist. If people help protect them, it is likely they will live on for many years to come.

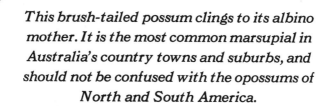

This brush-tailed possum clings to its albino mother. It is the most common marsupial in Australia's country towns and suburbs, and should not be confused with the opossums of North and South America.

The Amazing Kangaroo

When the Dutch Captain Pelsart's ship, *Batavia*, was wrecked off the coast of Australia in 1629, he and his crew made an amazing discovery. They spotted an animal unlike any they'd known in Europe. It moved around by leaping on its hind legs and sometimes carried a baby in its pouch. In this way, Europeans were first introduced to the kangaroo, although they didn't know what to call it at the time.

More than one hundred years later, in 1770, the famous English explorer Captain Cook was also shipwrecked near Australia. While his ship, *Endeavour*, was beached for repairs, he, too, spotted a strange animal. He asked the aborigines, or natives, what it was. "Kangaroo," they replied. And so Captain Cook introduced the first aboriginal word into the English language.

Today we know there are at least forty-five species, or kinds, of kangaroos. Like other marsupials, they live in Australia, New Guinea, and several nearby islands. Australians have given their native animals a nickname —*roo*. They've nicknamed kangaroo babies, too. These are called *joeys*.

Many people believe that all kangaroos are one size. Actually, they can be small, medium, or large. The tallest can grow to be 7 feet (2 m) and weigh more than 200 pounds (90 kg). That's taller and heavier than most humans. The smallest are the size of rabbits and weigh only about 2 pounds (0.9 kg).

Scientists call kangaroos *macropodids,* or "big-footed." The largest have feet 10 inches (25 cm) long or more. Those that are medium-sized and have smaller feet are called *wallabies.* These vary in size and looks, according to where they live. Some live in rocky cliffs, some in swamps, and others in open grasslands.

Dama wallabies, one of the smaller species, can drink seawater and survive.

Kangaroos living today are the product of millions of years of evolution. In one major respect, they still resemble their ancestors because they carry their young in a pouch. But in other ways, they have adapted, or changed, to meet the specific demands of their environment. Some scientists believe that kangaroos' teeth have changed to enable them to eat grass instead of meat. And, many kangaroos have grown to be faster and more powerful than their earliest ancestors. Some kangaroos can reach speeds of 40 mph (64 kph) for short sprints and are powerful enough to kill a dog with a single kick.

For years, kangaroos have intrigued scientists. How do they give birth? Why do some produce two types of milk? Why do they lick their inner arms? Gradually, their studies and experiments yielded fascinating answers.

Meanwhile, kangaroos have captured the imagination of other people. Circus trainers have taught them to box because of their unusual fighting skills. Cartoons feature kangaroos performing silly stunts, too. And in one award-winning children's film, *Me and You Kangaroo*, a pet kangaroo smashes china and furniture by leaping through its owner's home.

But despite heavy publicity, there are a number of facts that most people don't know about the kangaroo. Facts that are quite amazing.

We know that kangaroo mothers carry their babies, or joeys, in their pouches. But how do these babies get there? People believed that kangaroo babies were actually born in these pouches. Scientists have proved this isn't true. At birth, all marsupial newborn must make a journey to reach the pouch.

BIRTH Depending on the species, thirty-three to thirty-eight days after mating, female kangaroos give birth. The female searches for a quiet, safe place. There, she pokes her head inside her pouch and licks until it is spotless. Then she assumes what is called the *birth position*. She sits with her hind legs extended forward, her tail stretched between them.

In this position, the baby is born. It passes from her uterus into her fur. This tiny creature, shaped like a bean, is smaller than your little finger. It has no ears or eyes and is so pink it seems almost transparent. What it does have are tiny forearms equipped with sharp claws. These, and these alone, help it make its journey.

Moving its arms in a swimming motion, this tiny creature struggles through its mother's fur toward the pouch. The mother doesn't help. Instead, she watches or begins licking away the birth fluid. This action convinced many people that the mother actually licked a trail, or birth track, for the newborn to follow. Scientists know this is not so. The mother may occasionally smooth down a rough spot in the baby's path, but the newborn makes its journey unaided. The 6-inch (15-cm) journey doesn't take very long. In less than five minutes, the newborn enters its new home—the pouch.

THE POUCH The pouch is a fold of skin like a pocket on the mother's belly. For many months to come, it will provide food, shelter, and warmth. Now it is the newborn's crib or incubator. Once inside, the newborn tightly grasps one of its mother's four teats in its little mouth. The teats are the mother's mammary glands. Through the teat, the newborn receives the nourishment it needs to grow—milk.

The teat swells in the infant's mouth until it is so firmly attached it appears to be stuck. When people first examined the kangaroo's pouch and found this little creature practically glued to a nipple, they believed that babies were born in the pouch. That is what gave rise to the common misconception of pouch birth.

If scientists are very careful, they may remove the baby and return it without harming it. But the baby needs the mother's nipple. It is its lifeline. While in the pouch, the baby will grow and develop until it looks like a kangaroo.

The kangaroo's pouch is better developed than any other marsupial's. Some marsupials have hardly any pouch at all. The newborn just dangle from their

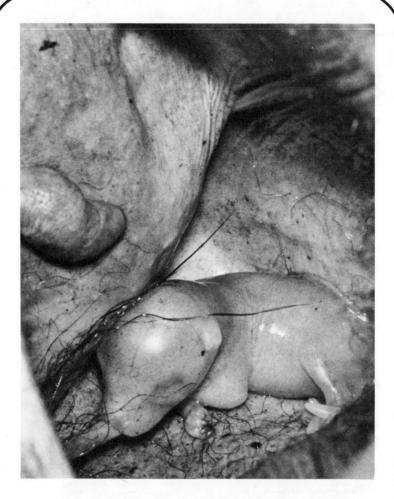

The mother's nipple remains inside a newborn kangaroo's mouth until the baby grows big enough to leave the pouch.

mother's nipples. The kangaroo has a forward-opening pouch, but some marsupials have a backward-opening one. A marsupial's habits usually determine which way the pouch opens.

GROWING UP Some species take longer to develop than others, but in approximately five months the joey pokes its head outside the pouch for the first time. Fur covers its head, but its ears still flop over. Although it is quite curious about the new world, the joey is also very cautious. Any sudden disturbance sends its head immediately back into the pouch. The joey has a chance to see many things, anyway. The pouch is a mobile home. Wherever mother goes, joey can peep its head out and look around.

JOEY'S FIRST STEPS. At six months, it's time for joey to try its new legs. Not all joeys are anxious to leave their snug homes. If joey protests, mother takes matters into her own hands. She bends over, tips her pouch, and joey tumbles out.

Gradually, the joey scrambles from the pouch on its own, but is still easily frightened. A sudden noise will send it scurrying back to the pouch. If there is no danger, the mother stands straight up to prevent the joey from climbing in. Sometimes the joey acts like an ostrich and buries its head in the pouch for comfort.

However, if the mother senses real danger, she reacts quite differently. She beckons her joey with soft clucking noises and allows it to dive headfirst into the pouch. Inside, the joey does a quick somersault until its head faces the pouch opening. Joey may be frightened, but doesn't want to miss any of the action.

A CAREFREE LIFE. During the next few months, joey leads a carefree life in and out of the pouch. In the morning, mother does the washing. Hugging joey closely to her to prevent any wriggling away, she gives joey a good bath with her tongue. She also cleans herself, concentrating especially on the pouch. Joey's mobile home is kept immaculate at all times.

The rest of the day is devoted to short trips in the pouch and playing. Like most young animals, joeys are extremely playful. They wrestle and frolic with everything—other joeys, plants, and their mothers. Mother is the biggest target for fun and games. A joey grapples with her ears, tugs at her fur, snips at her food, and generally doesn't give her any peace. Nevertheless, the mother is very patient and usually doesn't protest.

Adult males, on the other hand, have much quicker tempers. If a joey playfully approaches one, he might easily give the youngster a quick cuff to discourage any further nonsense. The joey soon learns that some kangaroos are not good playmates.

Joey's mobile home

JOEY AT HEEL. When joey reaches eight to ten months, it has grown too big and heavy for mother to carry. Each mother makes her own decision as to exactly when this is. Then the pouch is forbidden to the joey. It is permanently evicted and no amount of struggling or pleading will change mother's mind.

*Below: Joey stays close to mother.
Right: Joey-at-heel still
drinks from its mother's pouch.*

The joey need not feel totally abandoned, though. Mother still feeds joey from the pouch, and she is also very concerned about joey's welfare. The joey is kept closely by her side, and is referred to as a joey-at-heel or young-at-foot. The joey will complete the rest of its development in this way.

COMING OF AGE. At eighteen months, the joey achieves complete independence. Mother no longer provides food or protection. The young kangaroo is now capable of fending for itself.

Kangaroos living today are the product of millions of years of evolution. Some scientists believe that they have developed teeth suitable for grazing because grass is the most abundant food in the open interiors of Australia. If this is true, then it is a good example of how living creatures adapt, or change, to survive.

Today most species of kangaroos are called *herbivorous* because they eat only grass and other plants. People could also call them late-night snackers because they graze mostly at night.

Like other grazers, kangaroos have certain teeth that are arranged differently from the teeth of *carnivorous*, or meat-eating, animals.

THE TEETH The kangaroo has six upper front teeth and two long lower ones that jut forward in its mouth. An empty space called the *diastema* separates the front teeth from those in back. The back teeth are called *cheek teeth*.

Each section of the kangaroo's mouth plays an important role in grazing. The front teeth grasp blades of grass. The diastema helps keep the grass in place as the kangaroo continues to graze. And the cheek teeth grind down the grass. Sometimes a kangaroo will take a break during grazing. It rests back on its hind legs and tail, and long blades of grass dangle from its mouth. These blades of grass are kept in place by the diastema.

All grazing animals have similar teeth patterns. However, the kangaroo's mouth has one major difference.

Blades of grass dangle from this kangaroo mother's mouth. The diastema keeps them in place.

The two halves of its lower jaw can move against each other like scissor blades. This is possible because the jaw is not fused together by bone like that of other grazers. Instead, it is held together by bendable cartilaginous material. This unique feature is useful. It allows the kangaroo to chew many grass blades at once.

As a kangaroo ages, its teeth gradually fall out. If it lives to the age of twenty, chances are that few teeth will be left. Without teeth, the kangaroo cannot graze. Then it starves to death.

A kangaroo grazes on Australia's open grasslands.

Most marsupials run on all fours. Not the kangaroo! Only its powerful hind legs touch ground as it leaps from place to place. Its forepaws, or arms, are tucked in toward its chest, and its long, thick tail curves slightly upward behind.

Leaping is a fast mode of travel. In a single leap, some kangaroos can cover 37 feet (11 m), or approximately five times their own height. Most kangaroos like to cruise along at 12 to 15 mph (19 to 24 kph), but, if chased, some can speed along at 40 mph (64 kph) for short periods of time. Leaping is also a great help in scaling fences or other obstacles. Some kangaroos can easily sail across a 10-foot (3-m) fence.

How does the kangaroo make these fantastic leaps? Its feet, tail, and hind legs tell us.

THE BIG FEET A kangaroo's hind feet are like springboards. They enable it to leap high in the air. Each foot tapers into four toes. Only the middle toe and a shorter outer one help the kangaroo leap. The middle toe is much bigger than the others and ends in a sharp, solid, triangle-shaped nail.

On the inside of the foot, two tiny toes are joined by one piece of skin. Each little toe has its own curved nail. These toes are useless for leaping because they don't touch the ground. Instead, the kangaroo scratches and grooms itself with them.

Some kangaroos can leap higher than 10 feet (3 m).

The kangaroo's long muscular tail acts as a prop.

THE LONG TAIL A kangaroo's tail is long, thick, and muscular. During leaping, it sticks out behind, curves slightly upward, and acts as a balancer for the kangaroo's weight. When the kangaroo stands still, the tail also helps. It extends down, forming a tripod on which the kangaroo rests.

THE HIND LEGS A kangaroo's hind legs are Z-shaped, like those of frogs or other jumping animals. Three fourths of the kangaroo's bulk or weight resides in its legs, so it is no wonder that they are so powerful. The hind legs are also divided into three lengths—the thigh, the shin, and the foot. If your feet were as long as a kangaroo's, scientists might place you in the Macropodidae family along with them. *Macropodidae* comes from the Latin word meaning "great foot."

A kangaroo mob

Kangaroos live in groups called *mobs*. Many years ago, mobs sometimes consisted of a thousand kangaroos. Today, since the kangaroos' numbers have been greatly reduced by hunters, mobs are much smaller. People have reported seeing fifty kangaroos clustered together, but the average is ten to twenty.

MOBS Mobs consist of kangaroos of all ages and both sexes—males, females, and their joeys. The strongest and usually the largest male in the mob is called an "old man." Sometimes he bosses the others around. If a younger male challenges the old man to a fight and wins, he becomes the old man.

A mob is a loosely knit group. A kangaroo may stay with one mob for several years or move on to another after several days or weeks. The only real family unit is the female and her joey.

There is hardly any community spirit or cooperation in kangaroo mobs. Males do not protect females; the strong do not look out for the weak. When danger strikes, it is every kangaroo for itself. They scatter in all directions in a frantic effort to escape. In this respect, kangaroos differ from other animals that band together, like cattle or sheep. No one has to worry about being trampled in a stampede.

Although kangaroos don't take a common stand against enemies, they do have a way of warning against danger. One kangaroo may thump its foot against the ground, or else leap higher than usual. These movements tip off the others that something threatening may be lurking nearby.

MOB LIFE-STYLES

Kangaroos are much affected by weather. They have been known to travel great distances to find water or shade. But on very hot days, they really like to take it easy. If they can't find a shady tree, they dig shallow pits to lie in. Kangaroos usually lie on their sides, but they have been seen lying completely on their backs. A large kangaroo with all fours in the air is a pretty strange sight.

Even though lying down, kangaroos are always on the alert. Their ears act as antennae to pick up unusual sounds. Kangaroos don't have to turn their heads for better listening enjoyment. Their ears can turn 180 degrees from front to back.

Kangaroos also have another way of keeping cool in hot weather. Besides panting and sweating, they lick the insides of their arms. The evaporating moisture cools them down. Sometimes, even when kangaroos are leaping, they will stop to lick their inner arms. Scientists often wondered why kangaroos chose to lick this area, since it is so small. Studies conducted by T. J. Dawson and his colleagues gave them an answer. Since there was

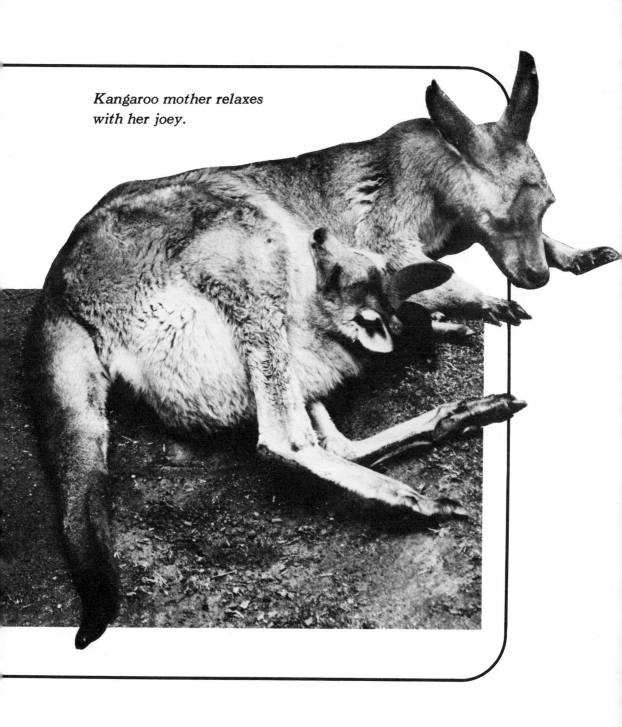

Kangaroo mother relaxes with her joey.

an intricate network of blood vessels in that area, great heat transfer resulted. Licking was an effective way of cooling down.

COURTSHIP Kangaroos usually mate at night. Since they band together in mobs, females rarely lack suitors. Sometimes more than one male becomes interested in a female. Then they fight. The winner becomes the female's partner for the night.

And one night it is. The male begins courtship by grazing next to the female. He may nuzzle her, and softly woo her with clucking noises. Finally he mounts her from the rear, and they mate. Afterward, they may remain together until sunup. At this time, the romance is over. Male and female go their own separate ways. Birth and child rearing are strictly the female's domain.

BREEDING At eighteen months, females are ready to mate and have babies. Kangaroo mothers never raise large families. Only one baby is born at a time. However, females breed until they die. Possibly this is nature's way of ensuring the survival of the breed.

On the other hand, males must wait until they are three to four years old to mate. Before this time, they may congregate in small groups. Peaceful grazing occupies most of their time, but occasionally they indulge in a little recreation—quick sparring matches.

Kangaroos aren't aggressive creatures, but they do fight. Sometimes it is nothing more than two young, frisky males testing their strength. Sparring matches are good practice in preparing for battles with their real enemy, the dingo, or wild dog. At other times, if two males are vying for a female, or if an old man is challenged, the fights may be more serious, sometimes resulting in grave injury or death.

Females, on the other hand, never fight. They are too busy bringing up joey.

A playful poke

WRESTLING A loud hacking cough is a kangaroo's way of starting a fight. Fighting kangaroos resemble wrestlers in a ring. They circle around each other, sniffing, and sizing up their opponent. As they rear back on their powerful hind legs, the match begins. They dance about, they parry, each waiting for a good opening in which to deliver his terrible kick. Just as in fighting the dingo, kicks are swiftly dealt with both legs, the tail acting as a brace for their massive weight.

The ripping nails on a kangaroo's hind legs are most damaging, but the sharply clawed forearms can be harmful, too. They slash at eyes and ears, which is why dueling kangaroos hold their heads far back. Sometimes kangaroos lock chest to chest in a clutch like human wrestlers. And sometimes one kangaroo manages to pick up his opponent and toss him through the air. This is quite a feat, since the sailing kangaroo may weigh 200 pounds (90 kg).

BOXING Because of their unique fighting abilities, kangaroos have been recruited to perform in circuses and sideshows. With boxing gloves fitted over their forepaws, they are actually trained to slug it out with humans. These matches are mainly for show, where neither contestant gets hurt. Because, if the kangaroo decided to use his powerful hind legs, the human would be in real trouble.

AN ODD MATCH. You wouldn't expect to find a kangaroo boxing a hippopotamus, but this once happened at the famous Hagenbeck Zoo in Hamburg, Germany. A red kangaroo, fitted with boxing gloves, leaped over a wall into a hippo's space and punched it in the nose. The huge animal was so surprised that it didn't try to hurt the kangaroo.

A kangaroo wearing boxing gloves in a playful clutch.

In nature, different types of animals sometimes form cooperative relationships to help each other. Some kangaroos enjoy such a bond with two fine feathered friends—the peewee and the willie wagtail. These birds rid the kangaroos of insects in exchange for food.

PEEWEES. These black and white birds hunt for dinner by walking around the kangaroo and pecking into its fur. The ticks and fleas that burrow into the kangaroo's body are pests the kangaroo is glad to lose.

WILLIE WAGTAILS. These birds are also black and white, but much more active than the peewees. They actually perch on the kangaroo's rump to dart after mosquitoes swarming above. Willie wagtails love mosquitoes and catch loads of them. During the nesting period, the willie wagtails receive another bonus. The kangaroo allows them to pull out bits of fur for nest-lining material.

CHANGING OF THE GUARD. The peewees and willie wagtails don't go far afield. They perform their pest patrol only when the kangaroo is in their territory. Once the kangaroo leaves, different peewees and willie wagtails take over.

Two nature author-photographers, Kay and Stanley Breeden, observed this happening on Bribie Island in Moreton Bay, Australia. They reported that the new birds taking over looked like a "changing of the guard."

SELF-SERVICE. Kangaroos do not rely solely on birds to ward off pesty insects. When they feel an itch, they scratch and comb through their fur with the "grooming claws" on their feet or with one of their five-clawed hands. Surprisingly, kangaroos have better control of their feet than of their hands and use them near sensitive areas like the eyes.

Thick kangaroo fur needs lots of grooming. Long nails are a big help.

The largest kangaroos have hind feet over 10 inches (25 cm) long. These are the big reds, the great grays, and the euros. When people think of kangaroos, they most likely have one of these in mind.

BIG REDS Big reds are not always red. Most females are a pretty shade of bluish-gray and are often called blue fliers because of their grace and speed. In some areas males are bluish-gray, and females are red.

Scientists have long been amazed by the fact that most species of female kangaroos can produce two types of milk at once. They aren't sure how their glands manage this, but closer study of big red females told them why. Within a few days after giving birth, the female mates again. An embryo, or the beginnings of a baby, is formed, but grows no bigger than one hundred cells. This embryo remains in the mother's uterus as kind of a reserve. By the time the first joey permanently leaves the pouch, the embryo has grown big enough to drop from the uterus into the fur. Then it travels to the pouch and attaches itself to a nipple like any other newborn. Only the milk it needs is different from the milk the first joey is still drinking. This is why the mother is equipped to produce two types. One nipple gives milk with a high sugar content for the newborn. Another gives high-protein and high-fat milk to the older joey.

An interesting change occurs in the male red's fur during mating season. Gland secretions cause a rose-red substance to form on his throat and chest. When the male smears this on his back with his forepaws, the fur turns bright red.

A big red takes off.

CLEVER KANGAROOS. For a long time, many people believed that kangaroos were rather stupid. A red kangaroo at Munster Zoo in Germany proved otherwise. D. H. Neumann trained a red kangaroo and an American opossum to tell the difference between pairs of drawings on paper by rewarding them with food for the right choice. He found that not only did the kangaroo choose correctly more times than the opossum, but it also had a better memory. The red kangaroo still made correct choices after five months. The opossum forgot after two weeks.

GREAT GRAY This large kangaroo is appropriately named. It grows heavier and more powerful than other kangaroos, and its fur is a silvery gray. When it is leaping, its black-tipped tail is said to go up and down like a pump handle. Great grays live in the forests of Australia and nearby islands, so they are also referred to as forester kangaroos.

The habits of great grays are similar to those of big reds, except that the female does not mate again until her first joey has permanently left her pouch. Therefore, great grays do not develop reserve embryos like the big reds and most other kangaroos. In addition, their joeys stay in the pouch longer than others.

Great grays, also called forester kangaroos, live near wooded areas.

Euros, or wallaroos, have thick, coarse fur.

EURO Scientists could not name this kangaroo by its shade, because its coarse fur varies from fawn to reddish-brown to dark gray. Occasionally all-black euros have been spotted, but usually just the forepaws and feet are dark.

Wallaroo, another name for these marsupials, is actually a combination of two names—*walla*by and kan*garoo*. These kangaroos are so named because they are smaller than great grays and big reds, but bigger than wallabies. They are also sometimes called hill kangaroos, since they live in Australia's rocky hills and gullies. In keeping with their rugged environment, euros have sturdy bodies, thick necks, and shorter legs than their bigger relatives. Rough soles on the bottoms of their feet help them negotiate rocky surfaces.

Euros are known for their ability to live without water for long periods of time. They manage this by hiding in caves to escape the sun's heat. Sometimes, when they grow very thirsty, they dig for water with their forepaws. They can dig as deep as 3 feet (1 m), and other animals often drink from their watering holes. In this way, euros sometimes help other animals.

There are over forty-five species, or kinds, of kangaroos. Those that are medium-sized and have feet smaller than 10 inches (25 cm) are called *wallabies*. And there is one kangaroo that makes its home in a different place from all the rest.

TREE KANGAROOS Tree kangaroos actually do live in trees. They are the aerialists of the breed, making great leaps between trees and long jumps to the ground. Some jump from tree limbs as high as 60 feet (18 m), their long tails balancing them along the way. If they climb down, it's tail first, like a fire fighter sliding down a pole.

Tree kangaroos are built differently from their ground-dwelling relatives. Their arms are better developed and almost as long as their legs. Rough pads on the bottoms of their feet cushion their landings. Their ears are shorter and more rounded than those of other kangaroos.

Tree kangaroos sleep with their heads tucked between their legs. Since their tree homes are mainly in rain forests, nature has provided them with a water-shedding device called a *whorl*. This is like a cowlick on the back of their necks. When it rains, the whorl prevents water from soaking into the fur. Instead, the rain slides off in all directions.

Tree kangaroos are built differently from their ground-dwelling relatives.

Aborigines on Australia's Cape York Peninsula are quite fond of the flesh of tree kangaroos. They train dingos to hunt and track them down. When the kangaroo scrambles up a tree, one hunter climbs after it. The kangaroo, nervous and frightened, jumps down. Unfortunately, it lands in the midst of a pack of dingos waiting below.

PRETTY-FACE WALLABY Originally named for its beauty, this marsupial has a silvery fur coat with a contrasting dark face banded with white. Another white stripe runs beneath the tail from hip to hip. But the prettyface has also been called the gray face, the blue flier, and the southern whiptail. This last name refers to a tail so long and slender that it equals the length of the wallaby's head and body. When the prettyface leaps, its tail curves upward, resembling a whip.

Even though their name changes, the prettyfaces' habits remain the same. They live near hilly areas and are thought to be more social than other kangaroos because they band together in larger mobs and stay together for most of the year.

NAIL-TAILS AND OTHER WALLABIES Other kinds of wallabies are also named for their physical characteristics. The *nail-tailed wallabies,* or *nail-tails,* are named for a small horny growth like a fingernail hidden in the tips of their furry tails. They are also

The pretty-face wallaby was named for its good looks.

Bennett's albino wallaby is not often seen in the wild.

Rat kangaroos gather nesting materials with their tails.

called "organ grinders" because of one odd habit. They extend their arms sideways and circle them as they leap.

Still other wallabies are named for their habitats. *Swamp wallabies* live in damp areas and gullies. And *rock wallabies* live in the caves of rocky hills and cliffs. These creatures are spectacular acrobats, ricocheting between ledges like bullets.

The *quokkas* aren't named for their looks or where they live. Instead, they have an island named after them. (Although they probably wouldn't like its name.)

In 1696 the Dutch navigator Willem De Vlamingh landed on an island near Western Australia. Because quokkas are so small, he thought the place was infested with rats. And so he called it Rottsnest, meaning "rat's nest."

RAT KANGAROOS Actually, some kangaroos are no bigger than rats. These are called the rat kangaroos, the largest of which are rabbit-sized. Rat kangaroos live in a variety of places: some in rain forests, others in dense undergrowth. One species, called *boodies,* actually burrows under the earth like moles. Boodies also have an unusual mode of combat. They lie on their sides to lash out with their hind legs.

As grazers, kangaroos never hunt other animals for food. But not all the creatures in their territory are as friendly. The kangaroo has several deadly enemies.

Joeys face the greatest danger. Their small size makes them an easy target for attack. Birds of prey, especially the wedge-tailed eagle and the white-breasted sea eagle, take advantage of this. They swoop down to attack joeys, but not the adults. Several kinds of goanna lizards also attack joeys. And foxes find them a satisfying meal.

DINGO

The dingo is the native Australian dog. Unlike house pets, dingos are wild and quite fierce. They attack kangaroos of all sizes.

One lone dingo may attack a kangaroo, but dingos usually prefer teamwork. Several members of a family group join forces to corner the kangaroo. In the past, the dingo was the kangaroo's greatest natural enemy.

THE KANGAROO FIGHTS BACK

Most kangaroos would rather flee than fight. But sometimes a large strong male chooses not to run. If only one dingo attacks, he may fight back.

The male's powerful hind legs are deadly weapons. When cornered, he strikes out with both at once. His steel-hard claws do great damage. A single kick can rip open a dingo's stomach and kill it.

The dingo, Australia's wild native dog, is the kangaroo's natural enemy.

An underwater shot of a kangaroo swimming. Kangaroos often dive into the water to escape their enemies.

The kangaroo also uses his forearms to defend himself. He may clutch the dingo tightly to his chest while delivering his savage kick.

Sometimes the kangaroo prefers to outsmart the dingo. If a waterhole or creek is nearby, the kangaroo leaps in. Kangaroos are excellent swimmers. If the dingo pursues, the kangaroo stops and waits. When the dingo is within reach, the kangaroo holds his enemy under water with his forearms until it drowns.

John Gilbert, one of Australia's naturalists in the early 1800s, reported an interesting eye-witness account. He watched a kangaroo drown a dingo in shallow water. The kangaroo held the wild dog under with his foot.

Today the dingo poses a much lesser threat. Since the dingo also hunts and kills sheep, people went to great lengths to eliminate it. The government offered high bonuses to people for bringing in dingo scalps. Some people made their living strictly from dingo hunting, or "dogging." This practice greatly reduced their numbers. However, whenever people tamper with nature's balance, other problems often arise.

PEOPLE Most Australians are very fond of their native kangaroo. They are proud that the kangaroo is peculiar to their own homeland, but also they like its speed, grace, and friendliness. Some Australians even feel that the kangaroo's great leap symbolizes the forward thrust of their nation. Drawings or photographs of kangaroos are imprinted on their national coat of arms, aircraft, postage stamps, and even coins and bank notes. Their international rugby team is called the Kangaroos.

Not all Australians, however, love the kangaroo. Sheep ranchers are not at all pleased by their presence. Big business has killed them for fur and pet food. And motorists often find them a menace.

SHEEP RANCHERS   When people settled in Australia, the kangaroo was in danger. Men hunted it. Much forest land was cleared to build cities and industries. Where would the kangaroo find food?

Sheep ranchers unwittingly solved part of this problem. The land they cleared for sheep grazing was perfect for the kangaroo. Also, the wells they dug for their sheep provided wonderful drinking fountains during dry spells, or droughts.

However, when the sheep rancher discovered kangaroos grazing on his pastures, he became furious. He felt the kangaroo was consuming too much valuable food. Many sheep ranchers, therefore, shot the kangaroo. The sheep rancher became the kangaroo's enemy.

Today, government law forbids sheep ranchers to shoot kangaroos. If a sheep rancher has a serious problem, he must call the government authorities.

A young visitor at the zoo makes friends with twin joeys. Twins are quite rare.

The government sends men to survey the area by helicopter. If there are too many kangaroos in a specific area, some will be killed. Killing kangaroos for purposes of population control is called *culling*.

MOTORISTS Kangaroos never look before they leap. They charge into the road without any warning. This often endangers motorists' lives and their own. Slamming into a 200-pound (90-kg) kangaroo at 50 mph (80 kph) is no joke.

In areas where there is a heavy concentration of kangaroos, the government puts up road signs. They read: KANGAROO CROSSING.

People who live in these areas have another method of dealing with this danger. Their cars are equipped with special bumpers. These bumpers offer some protection to their cars. Unfortunately, they don't help the stricken kangaroo.

BIG BUSINESS For many years, the word *kangaroo* spelled big business to some people. Thousands were slaughtered for their furs, skins, and meat. Leather products were fashioned from their hides and exported all over the world. Coats, rugs, and souvenirs were made from their rich, soft fur. Their furs were also used to produce imitations of the koala, a marsupial wholly protected by law.

Kangaroo meat was turned into pet food. People didn't find it too tasty, but it was exported for sausage making, and some gourmets considered kangaroo-tail soup a delicacy.

Fortunately, many objected to this widespread kangaroo killing for profit. Not only Australians protested, but private citizens and conservation groups from all over the world. Finally, the Australian government took action. Today no endangered species of kangaroo is allowed to be killed.

Mother and child

Nearly 175 marsupials live in the Australian region, but the koala is probably the best loved. Perhaps you have never seen a live koala, but you may have watched one on television. One Australian airline uses these adorable creatures in their commercials.

Why are koalas so popular? Possibly because they're harmless, trusting, and look so much like cuddly teddy bears. Their bodies are covered with thick gray fur, their eyes look like round buttons, and their noses appear to be made of black patent leather. Koalas are also small enough for humans to pick up and hold. The largest stand only 2 feet (60 cm) high and weigh roughly between 18 and 25 pounds.

Koalas eat only one kind of food—the leaves of eucalyptus trees. Some people even think koalas smell like cough drops because eucalyptus oil is used in making this candy medicine.

John Price was the first European to describe the koala, in 1798. At that time the natives called this animal by several names—cullawine, colo, and koala. But the name that stuck was koala, meaning "drinks no water." Actually, sometimes koalas do drink water, but mostly they obtain the fluids they need from the juicy eucalyptus leaves.

Even though koalas eat only eucalyptus leaves, they select their meals very carefully. Before dining, they take a deep sniff. This is because there are over five

Koalas are light enough for humans to hold.

A koala nibbles on its only food—leaves from the eucalyptus tree.

Koalas are built for a tree-dwelling life.

hundred varieties of eucalyptus trees, and during certain times of the year some give off poisonous prussic acid. The koala's nose tells it which to avoid.

Once the koala make its selection, it happily munches away. Koalas can consume up to 2½ pounds (just over 1 kg) at one time. They may use their cheek pouches to store food temporarily, but some people have spotted koalas dozing off in the middle of a meal. The eucalyptus leaf remains firmly clasped between their teeth.

Like most other marsupials, koalas are nocturnal creatures. This means they spend most of their waking hours at night. During the day, koalas wedge themselves into the forks of eucalyptus trees to snooze. These forks make good sleeping places because they are high above ground and are less likely to break than other branches.

Koalas are also great tree climbers. The first two fingers on their hands are opposed to the other three. On their feet, the first toe is opposable to the rest. This enables them to grasp branches very much as we use our thumbs. Also, their sharp claws give them a good grip. However, on the ground, koalas are quite slow and clumsy. If another animal attacks, they can only protect themselves by scurrying up a tree.

Professor W. A. Osborne has explained the reason for this. According to his studies, the koala's powerful thigh muscle is attached much lower on the leg than other animals'. This means that the koala's body is constructed for climbing, rather than running.

BRINGING UP BABY

The female koala gives birth to one baby at a time. Like other marsupials, the young completes development in the pouch and is nourished by milk from the mammary glands. But unlike the kangaroo's, the koala's pouch opens backward. For a long time, scientists puzzled over this. Then they discovered the rear opening served an effective, if unappetizing, purpose. During weaning, while her young is still in the pouch, the mother releases a yellowish green slime, composed of partially digested eucalyptus leaves, from her anus. By licking this substance, the baby gradually learns to digest the food it must eat to stay alive.

After around six months, the baby koala leaves the pouch. It still hasn't grown large enough to take care of itself, so now it rides piggyback on its mother. People have been surprised to watch mothers discipline their young. They spank them. But koalas show great affection, too. They hug their children or rest them on their laps.

At one year, the koala is big enough to live on its own.

Koala mothers carry their young piggyback style.

Left: a koala mother shows love by hugging her child. Right: a koala naps in the fork of a eucalyptus tree.

Many koalas live in government preserves.

STRANGE KOALA SOUNDS

In spite of the koalas' friendly appearance, they mostly lead solitary lives. Only during the mating season do they become more intimate. Then a male sometimes collects a harem of two to three females.

Males also become much noiser during breeding times. They call to their mates in harsh voices that people say resemble a handsaw grating through a thin board. Perhaps this explains a woodcutter's strange account of a koala attack. This man reported to naturalist B. E. Carthew that a large male bit his leg and wrist while he was working with a chain saw. The koala very likely thought the chain-saw noise was another male's challenge. Nevertheless, this is an extraordinary account because koalas are considered quite harmless and rarely fight among themselves. During mating seasons, they usually growl to discourage other males.

If koalas are hurt or in danger, their voices change considerably. Then they wail. The sound of a hurt koala is particularly horrible to many people because it resembles the cry of a human child.

KOALA PETS

Before laws were passed to protect koalas, some people tried to keep them as pets. This had mostly disastrous results, for without fresh eucalyptus leaves the koala would die. However, there have been accounts of injured koalas surviving on bread and milk, some even developing a taste for jam. And, during World War I, one army unit, sailing for Egypt, kept a koala as a mascot. On board, they fed it apples soaked in eucalyptus oil, but the koala still became ill. Fortunately, this koala was eventually saved. Upon landing, it was once again fed its natural food—leaves from Egypt's eucalyptus trees.

THE KOALA'S ENEMIES Dingos, owls, and goanna lizards are all enemies of the koala, but for many years the koala's greatest enemies were people. Hunters killed them and sold their soft fur for high prices. People who called themselves "sportsmen" shot them because they made easy targets. Apparently these people weren't aware that it isn't very sporting to kill defenseless animals. People killed so many that the entire koala population was almost wiped out.

A SECOND CHANCE Fortunately, Australia finally passed strict laws to protect the koala. But because so many were killed, great efforts had to be made to preserve them.

During the 1930s, Australian children immediately joined the battle to save koalas. They grew different varieties of gum trees (eucalyptus trees) from seedlings supplied by the government. Later, these young trees were transplanted to preserves. The children also helped conservationists by taking koala surveys in Victoria, New South Wales, and Queensland.

Koalas are lassoed from trees for transfer to preserves.

Peru Public Library
Peru, Indiana 46970

Eventually, these efforts paid off. Today many koalas live in special preserves. People transfer koalas to their new homes by lassoing them from trees and catching them in giant nets. And signs have been posted that read: KOALAS CROSS HERE AT NIGHT. It is hoped that if people continue to help the koala, Australia's living teddy bears will be with us for a long time.

Other Marsupials

MARSUPIAL MOLES

Most people have heard of blind little creatures called moles. But did you know that some moles have pouches? They are called marsupial moles and live in Australia.

Marsupial moles are excellent examples of what scientists call *parallel evolution*. This means that biologically different animals grow up looking alike under similar environmental conditions. Marsupial moles look very much like the true moles of Africa. However, true moles are pouchless, and they are not related. Other Australian marsupials resemble animals we know. Pouched cats, wolves, and mice are some of them.

Marsupial moles are only 6 inches (15 cm) long. That's smaller than a kitten. They have no external eyes, and little holes replace their ears. A horny shield covers their noses. Their rich, silky fur varies from cream to a deep orange. When they are alive, the fur is so luminous it's almost like velvet. Marsupial moles don't have much of a tail. It is short and leathery and covered with rings.

Marsupial moles are ideally built for burrowing—which is exactly what they do. Two curved front claws scoop out sand or soil like shovels. They don't make permanent tunnels, though. As they dig, the soil falls behind them. People know when marsupial moles are underground because there are slight cracks or move-

A horny shield hides the marsupial mole's nose.

The opossum's young snuggle in her fur. An opossum has fifty teeth—more than any mammal in North America.

ments in the earth. Sometimes these marsupials surface for a breath of air. They don't stay long because they don't travel well above ground.

Scientists have observed the burrowing habits of these creatures in captivity. They act somewhat like nervous wrecks. They madly tunnel through the earth as if they were starving. When they find the earthworms or insects they like to eat, they instantly gobble them up. Then they rush about for more, but without a moment's notice they may drop off to sleep. When they awake, it's back to business as usual.

Not much is known about these marsupials' breeding habits. They do have pouches, which open backward, and bear one young at a time.

OPOSSUMS Opossums are famous for one remarkable trick. When in serious danger, they play dead, or as the expression goes, "play 'possum." They fall to one side, their tongues hang out, and their eyes stare glassily ahead. Another animal may poke and paw them, but nothing forces them to budge. Since most animals prefer to kill the animals they eat themselves, playing 'possum often saves these tricksters' lives.

Opossums are the only marsupials found outside the Australian region. Also called Virginia opossums, they were originally found in the southern states of America. Later they spread to New York State and New England

and are now found in many western states and Canada, too.

A grown opossum is roughly the size of a house cat, and its white face has a long snout. Its body is covered with thick grayish fur. Coarse dark-tipped hairs are scattered throughout. These are called *guard hairs*. Only the tail is hairless. It is long and scaly like a rat's and is often used as a fifth hand.

An opossum can twist its tail around tree limbs, then swing from it like a monkey. The tail also comes in handy at nesting time. The female collects twigs and leaves with her mouth, passes them to her hind feet with her front paws, and loops her tail to carry this material in a bundle.

An opossum has fifty teeth—more than any other mammal in North America. These teeth make a wide choice in diet possible. The opossum dines on persimmons and berries, birds' eggs, insects, snakes, and mice.

Opossum feet have five toes. All are sharply clawed, except the first toe on each hind leg, which is opposable, like a thumb. This makes opossums good tree climbers. When they walk through the swampy areas where they often live, their claws leave starlike tracks.

Like most other marsupials, the female has a pouch. She gives birth to big litters, sometimes as many as twenty-four. These newborn are so tiny you could fit all of them in a teaspoon.

The newborn must journey to their mother's pouch. Not all will survive. Inside the pouch are only thirteen nipples, arranged in a horseshoe shape. Those left without a nipple automatically die.

The newborn remain firmly attached to the nipples until they are approximately nine weeks old. During this period, the mother may take a swim. No baby drowns, because a muscle around the pouch's outer rim puckers closed.

When the babies are about the size of mice, they crawl out and climb onto their mother's back. Clinging to her fur with their claws, they ride along as the mother goes about her business. When they grow sleepy, some crawl back into the pouch. Others snuggle in her warm fur.

When the young are thirteen or fourteen weeks old, they leave their mother. Then they go off and find a tree hollow or hole all their own.

Further Reading

Breeden, Stanley and Kay. *The Life of the Kangaroo.* Sydney: Angus and Robertson, 1966.
Coerr, Eleanor. *Biography of a Kangaroo.* New York: G. P. Putnam's Sons, 1976.
Darling, Louis. *Kangaroos and Other Animals with Pockets.* New York: Morrow, 1958.
Eberle, Irmengarde. *Koalas Live Here.* Garden City, N.Y.: Doubleday, 1967.
Gould, *Marsupials and Monotremes (Australian).* London: Macmillan, 1977.
Hurd, Edith Thacher. *The Mother Kangaroo.* Boston: Little, Brown, 1976.
Jenkins, Marie. *Kangaroos, Opossums, and Other Marsupials.* New York: Holiday House, 1975.
Kohn, Bernice. *Marvelous Mammals: Monotremes and Marsupials.* Englewood Cliffs, N.J.: Prentice-Hall, 1964.
Lauber, Patricia. *The Surprising Kangaroo and Other Pouched Mammals.* New York: Random House, 1965.
Serventy, Vincent and Carol. *The Koala.* New York: E. P. Dutton, 1975.
Animals of Australasia. "World of Wildlife Series." London: Orbis, 1977.
Living World of Animals. London: Reader's Digest Press, 1977.

Index

Australia, 3, 4, 7, 34, 43

Baby kangaroos. *See* Joeys
Batavia (ship), 7
Big business, kangaroos and, 54–55
Big reds, 35–37
Birth
    of kangaroos, 10–11
    of koalas, 64
Blue flier wallabies, 35, 43
Boodies, 47
Boxing, 31–32
Breeden, Kay, 34
Breeden, Stanley, 34
Breeding, 28–29
Brisbie Island, Moreton Bay, Australia, 34

Cape York Peninsula, Australia, 43
Carthew, B. E., 69
Cheek teeth, 18
Cook, James, 7

Courtship of kangaroos, 28
Culling, 54

Dawson, T. J., 26
Diastema, 18
Dingos, 29, 43, 48, 51, 70

Ears of kangaroos, 26
Endeavour (ship), 7
Enemies
    of kangaroos, 48, 51
    of koalas, 70
Eucalyptus trees, 59, 63, 70
Euros, 40

Feet of kangaroos, 21
Fighting, 28–29, 31, 48, 51
Food
    of kangaroos, 18, 20
    of koalas, 59, 63, 69–70
    of opossums, 78
Forester kangaroos, 37
Foxes, 48

Gilbert, John, 51
Goanna lizards, 48, 70
Gray face wallabies, 43
Grazing, 18, 20
Great grays, 37
Grooming, 33–34
Growing up
    of kangaroos, 13–17
    of koalas, 64

Hagenbeck Zoo, Germany, 32
Hill kangaroos, 40
Hind legs of kangaroos, 23

Intelligence, 37

Joey-at-heel (young-at-foot), 17
Joeys, 7, 13–17, 48

Kangaroos, 7–55
    big business and, 54–55
    big reds, 35–37
    birth, 10–11
    breeding, 28–29
    climate and, 26
    courtship, 28
    culling, 54
    ears, 26
    enemies, 48, 51
    euros, 40
    feet, 21
    fighting, 28–29, 31–32, 48, 51
    forester, 37
    grazing, 18, 20
    great grays, 37
    grooming, 33–34
    growing up, 13–17
    habitat, 8
    hill, 40
    hind legs, 23
    intelligence, 37
    leaping, 21, 23
    licking, 26, 28
    mating, 28, 35, 37
    mobs, 25–26
    motorists and, 54
    nail-tailed wallabies, 43, 47
    pouch, 3, 9, 10, 11–13, 16
    prettyface wallabies, 43
    quokkas, 47
    rat, 47
    rock wallabies, 47
    sheep ranchers and, 52
    size, 7–8, 35
    southern whiptail wallabies, 43
    species, 7, 41, 43, 47
    speed, 9
    swamp wallabies, 47
    tails, 23
    teeth, 9, 18, 20
    tree, 41, 43
Koalas, 3, 57–72
    birth, 64
    climbing, 63–64

Koalas *(continued)*
    enemies, 70
    food, 59, 63, 69–70
    growing up, 64
    as pets, 69–70
    popularity, 59
    pouch, 64
    protection of, 54, 70, 72
    sounds, 69

Leaping of kangaroos, 21, 23
Licking, 26, 28

Macropodidae, 8, 23
Marsupial moles, 74, 77
Marsupials
    defined, 3
    environment and, 4–5
    newborn, 3
    See also Kangaroos; Koalas; Marsupial moles; Opossums
Mating of kangaroos, 28, 35, 37
Milk, kangaroo, 35
Mobs, 25–26
Moles, 74
Motorists, 54
Munster Zoo, Germany, 37

Nail-tailed wallabies, 43, 47
Neumann, D. H., 37
New Guinea, 3, 7

Opossums, 3, 4, 37, 77–79
Osborne, W. A., 63
Owls, 70

Parallel evolution, 74
Peewees, 33–34
Pelsart, Captain, 7
Placenta, 3
Pouch
    of kangaroos, 3, 9, 10, 11–13, 16
    of koalas, 64
    of marsupial moles, 74, 77
    of opossums, 78–79
Prettyface wallabies, 43
Price, John, 59
Protection of koalas, 54, 70, 72

Quokkas, 47

Rat kangaroos, 47
Reserve embryos, 35, 37
Rock wallabies, 47
Roo, 7
Rottsnest Island, 47

Sheep ranchers, 52
Size of kangaroos, 7–8, 35
Sound of koalas, 69
Southern whiptail wallabies, 43
Speed of kangaroos, 9
Sugar glider, 3
Swamp wallabies, 47

Tails of kangaroos, 23
Teeth
    of kangaroos, 9, 18, 20
    of opossums, 78
Toes of opossums, 78
Tree kangaroos, 41, 43

Wallabies, 8, 35, 41, 43, 47

Wallaroo, 40
Wedge-tailed eagle, 48
White-breasted sea eagle, 48
Whorl, 41
Willie wagtails, 33–34
Wrestling, 31

You and Me and Kangaroo (film), 9

About the Author

Ellen Rabinowich has had a varied career in the arts. In addition to writing juvenile fiction and nonfiction, she has acted professionally and produced a one-hour dramatic film as well as promotional items for major corporations. She is the author of *Toni's Crowd*, a novel in the Franklin Watts Triumph series. This is her first book about animals.